P9-BJB-387

21ST CENTURY DEBATES

TRANSPORTATION

OUR IMPACT ON THE PLANET

ROB BOWDEN

Chicago, Illinois

21st Century Debates Series

Genetics • Surveillance • The Internet • The Media • Artificial Intelligence • Climate Change • Energy • Rain Forests • Waste, Recycling, and Reuse • Endangered Species • Air Pollution • An Overcrowded World? • Food Supply • Water Supply • World Health • Global Debt • New Religious Movements • The Drug Trade • Terrorism • Racism • Violence in Society • Tourism • Globalization

Published by Raintree, a division of Reed Elsevier Inc., Chicago, Illinois

For information, address the publisher:
Raintree, 100 N. LaSalle, Suite 1200, Chicago, IL 60602

Printed in Hong Kong
08 07 06 05 04
10 9 8 7 6 5 4 3 2 1

Library of Congress Cataloging-in-Publication Data

Bowden, Rob.
Transportation / Rob Bowden.
p. cm. -- (21st century debates)
Summary: Looks at the effects that various means of transportation have on ecology throughout the world, and what changes need to be made to expand the advantages of modern transportation and preserve the natural environment.
Includes bibliographical references and index.
ISBN 0-7398-5505-0 (Library Binding-hardcover)
1. Transportation--Environmental aspects--Juvenile literature. [1. Transportation--Environmental aspects. 2. Environmental protection.] I. Title. II. Series.
HE147.65.B68 2004
388--dc22

2003012733

Picture acknowledgments: Ballard Power Systems 21; EASI-ER 24, 29 and 51 (Rob Bowden); Chris Fairclough Photography 16, 22, 27, 30, 46, 47; HWPL 8 (Paul Kenward), 9, 10, 20 (Rolando Pujol), 36; Popperfoto 7 (Fabrizio Bensch), 12, 13 (Ian Hodgson), 23 (Larry Chan); Rex Features 39, 48, 58; Rex Interstock Ltd 25 (Wildtrack Media); Still Pictures 14 (NRSC), 17 (Dylan Garcia), 18 (Mark Edwards), 31 (Thomas Raupach), 50 (John Maier), 52 (Alex S. Maclean), 53 (Martin Bond); Topham 37, 42, 44, 57; Travel Ink 32 (Allan Hartley), 35 (Derek Allan); WTPix 4, 6, 11, 28, 33 and cover foreground, 41, 43, 49 and cover background, 54, 55.

Cover: foreground picture shows a light rail system in Sydney, Australia; background picture shows safe storage for bicycles in Ravenna, Italy.

Every effort has been made to trace copyright holders. However, the publishers apologize for any unintentional omissions and would be pleased in such cases to add an acknowledgment in any future editions.

CONTENTS

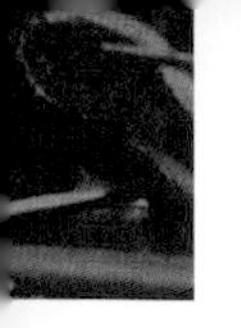

A WORLD ON THE MOVE

"We live in a world where international trade and travel are critical for economic success. They are also, very simply, a fact of life. Transportation and technology ... allow us to be more and more interconnected, and more productive."

Rick Kowalewski, Deputy Director, U.S. Bureau of Transportation Statistics, 2003

Stop to think!

If you were to step outside your home or school and stand still, even for a minute, you would notice that the world around you was on the move. You might

Whether on foot, bicycle, motorcycle, car, or bus, people and goods are constantly on the move at this busy intersection in Bangalore, India.

see a pedestrian or cyclist pass by, or cars and buses on the roads. You might hear a train on a nearby track or see an airplane flying overhead. The fact is that people and goods are moving all around us, 24 hours a day. We are so used to this movement that many of us give it little thought. Yet there are many reasons why we should.

FACT

More than 96 percent of the world's transportation is dependent on oil.

The benefits of mobility

Movement, or mobility, is not a problem. The mobility of people and goods is essential to the functioning of a country's economy. How would a city manage without its daily flow of commuting workers, or a factory without its delivery of raw materials and shipping of finished goods? Where mobility is inefficient, economies can suffer. Poor infrastructure (roads, railroads, etc.) in parts of Africa and Asia deters businesses from investing in the local economy because it is harder, and often more expensive, to move goods and people around.

Mobility is also socially important. It brings people together to meet, discuss, trade, relax, and enjoy themselves. It also gives people greater personal freedom, such as the ability to travel. These benefits are seen by many people in more developed countries as basic rights. Any threat to personal mobility is taken very seriously.

The costs of mobility

Unfortunately the benefits of mobility come with costs. And the costs are all linked to the types of transportation that make mobility possible. Apart from walking and bicycling, virtually all forms of transportation used today are dependent on oil as their energy source. However, oil is a nonrenewable resource and, when burned (to release its energy), it sends harmful emissions into Earth's atmosphere. This presents serious issues for both people and the environment.

VIEWPOINT

"Transportation is essential for moving people and goods, but it also has a broader role. It shapes our cities, stimulates economic growth, and makes possible ...[social] interactions."
Daniel Roos, Massachusetts Institute of Technology (MIT)

VIEWPOINT

"There has been a growing international [agreement] that prudence requires us to reduce the amount of CO_2 added to the atmosphere through human activities, including [transportation]."

World Business Council for Sustainable Development, "Mobility 2001"

Heating up

The biggest problem associated with transportation is the threat of climate change caused by the heating up of Earth's atmosphere. This heating, or "global warming," is caused by a buildup of certain gases in the atmosphere that act like a greenhouse, allowing the energy of the sun to reach Earth and then trapping its heat. The most important greenhouse gas is carbon dioxide (CO_2) which is released when fossil fuels, such as oil, are burned. Although CO_2 is released from numerous sources, transportation is to blame for an increasingly large proportion of emissions. In 1999 transportation was estimated to account for around 26 percent of CO_2 emissions from human activities.

As car use increases so does traffic congestion, here shown in Shanghai, China.

A growing concern

Emissions of CO_2 from transportation are certain to increase in the future as the number of conventional motor vehicles on the world's roads continues to grow. Already in 2001 there were an estimated 700 million motor vehicles. With around 40 million new vehicles being produced each year, the number on the world's roads is expected to reach 1.1 billion by 2020. Air travel worldwide is also increasing rapidly and is expected to double between 1995 and 2015. This will add to the CO_2 in the atmosphere and increase the threat of climate change.

Warning signs

Many experts believe we are already seeing the first signs of climate change, with more frequent abnormal weather events such as droughts and floods. Glacier fields and the polar ice caps are also melting at an alarming rate due to an increase in average global temperatures. If such trends continue, whole regions of the world could be seriously threatened by rising sea levels as the oceans warm and swell (water increases in volume as it gets warmer). Ironically, such changes could threaten the very mobility that is partly to blame for causing them. The challenge that lies ahead is to find ways of preserving mobility without harming the environment for future generations. This will demand a reassessment of transportation options for the 21st century. This book shares some of the ideas and opinions being expressed by the different contributors to this important debate.

Cars and trucks in the U.S. consume about 11 percent of the world's annual oil production, but the U.S. accounts for just 5 percent of world population.

Before reading on, think about your own use of transportation and how it affects others and the environment. See how many of those issues come up in the rest of the book.

Rescue workers navigate the flooded historic streets of Meissen, near Dresden in Germany, after the Elbe River flooded in 2002. Experts believe that global warming, caused partly by emissions from transportation, could lead to similar scenes in many countries in the future.

REVOLUTIONS IN TRANSPORTATION

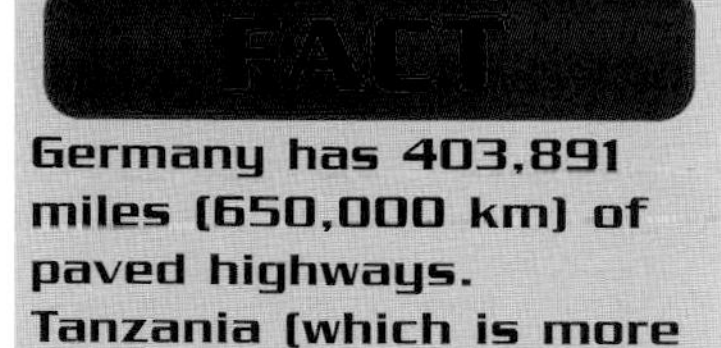

Germany has 403,891 miles (650,000 km) of paved highways. Tanzania (which is more than twice the size of Germany) has just 2,640 miles (4,250 km) of paved highways.

Continual change

The history of transportation is one of successive revolutions that have continually transformed the mobility of people and goods. These revolutions have shaped much of human history and have determined the nature and location of human settlements and economies. Take the ancient Egyptians, for example. It was their ability to build feluccas (an Egyptian-style sailing boat), for use on the Nile River, that allowed them to transport the granite and other materials they used to construct their magnificent temples.

This donkey performs a vital role, carrying water in Kenya.

In more recent times there has been a revolutionary move away from walking—our natural form of motion—to modern forms of transportation such as the car and the airplane. And yet, despite incredible progress in transportation, people in many parts of the world continue to depend on forms of mobility that have changed little for thousands of years. In Kenya, for instance, donkeys are still widely used to move goods between farmers' fields and the markets. Similarly, camel trains still remain a vital form of transportation in desert regions, such as those in North Africa.

A world of options?

In the early 21st century, it might seem that we live in a world full of transportation options. But in reality, those options are not equally available

In the hostile environment of the Egyptian desert, a camel is often a more reliable form of transportation than the modern car.

to all. The availability and use of transportation is closely related to levels of economic development. One way to look at this relationship is to consider the number of cars per person in different regions or countries. In the industrialized nations of North America, Western Europe, and Japan, for example, there is today one car for every two or three people. In less-developed countries, such as India and China, there is only about one car for every 250 people. Similar inequalities in transportation can be seen in other sectors, too. For example, Great Britain has a railroad network almost five times bigger than that of Nigeria, even though Britain is only about a quarter of Nigeria's size, and has less than half the population of Nigeria.

VIEWPOINTS

"A community without roads does not have a way out."

A poor man, Juncal, Ecuador, in World Bank, Voices of the Poor *(2000)*

"The relationship between [transportation] and development is neither obvious nor simple. Simply building a road does not guarantee that development will take place."

Department For International Development (DFID), Britain

VIEWPOINTS

"I was dreadfully frightened before the train started. In the nervous state I was in, it seemed to me certain that I should faint from the impossibility of getting the horrid thing stopped."
Thomas Carlyle, Scottish historian, describing a journey on the Liverpool & Manchester Railroad, 1842

"I began to experience an inexhaustible wonder at the gracious beauties of the world outside factory-land, and this sensation has never wholly left me. That first long [railroad] journey was as wonderful to me as if I had been riding upon the magic carpet in the Arabian Nights."
Joseph Robert Clynes, trade unionist and writer, describing early train travel in Britain, 1892

Transportation and the economy

Improving transportation in less-developed countries is essential to their economic development. In an increasingly competitive global economy, less developed countries need efficient and reliable transportation to attract foreign investment. This is nothing new. Throughout history economies have been highly dependent on transportation for their success. The Industrial Revolution in 19th-century Britain, for example, was partly driven by improvements in transportation, first with the development of the canal network and later the railroads. These allowed raw materials and finished goods to be moved around the country more easily. Factories were located close to these transportation networks and workers soon followed to take advantage of the new job opportunities, turning towns into cities. As these industrial cities grew, so did the demand for transportation.

A train leaves Paddington Station, London, in 1892. At the time this kind of transportation was considered revolutionary.

Overground, underground

With so many people in one location, it became cost-effective to run new forms of transportation such as streetcars (or trams, as they are known in

A streetcar (or tram) running on the streets of Toronto in Canada. Although dating back almost 150 years, streetcars are now making a comeback in many cities.

Europe). The first horse-drawn streetcars began operating in London in 1860, but it was the invention of the dynamo (electrical generator) in 1869 that led to a sudden growth in streetcar systems that could now be powered by electricity. By 1910 most cities in Europe and North America had developed streetcar systems. They became the main form of urban transportation and, as the cities expanded into the suburbs, so did the streetcars.

However, streetcars could only use a fixed route because of the wires, tracks, or cables needed to power them. The invention of the gas engine solved this problem and provided more versatile transportation. By the 1940s buses had replaced many streetcars and the old tracks and cables were gradually removed. In some places, however, new fixed-route transportation systems took the place of streetcars. Among the most important was the subway system. The subway used space efficiently by running trains in tunnels beneath the city instead of using up valuable surface land. New York, London, Paris, Boston, and Budapest all had subway systems by 1905, and many other cities soon followed their lead, including Chicago, Tokyo, and Moscow.

The first subway system in London was just 4 miles (6 kilometers) long and used steam-powered trains. It opened in 1863 and carried 9.5 million people in its first year.

VIEWPOINT

"In the 20th century the automobile provided tremendous benefits and raised the standard of living in much of the world. But it also had a major negative impact on the environment."
Bill Ford, Chairman of the Board, Ford Motor Company

Personal transportation

Early transportation systems were developed to serve industry and, later, to transport workers from their homes to the factories and offices where they worked. As incomes increased and people began to enjoy more leisure time, a new demand for personal transportation began to emerge. The railroads were one of the first ways that people enjoyed transportation for personal reasons. People began to take to the trains to escape the overcrowding and pollution of the cities. By the 1930s, special day-trip services were taking people from large cities to resorts in the country and along the coast.

Freedom of the road

The real revolution in personal transportation came with the development of the private car. Early models were expensive, but the development of mass production techniques led to cheaper cars, such as Ford's Model-T, which many average-income people could afford. Ever since the introduction of the Model-T in 1908, private driving has grown rapidly and is today the dominant form of transportation in the more developed regions of the world. The freedom that a private vehicle gave people has been the key to its success. Today many people living in more developed countries consider this freedom to be a basic right, but for the vast majority living in less-developed countries it remains beyond their reach.

Cars being manufactured in a factory in Detroit, Michigan, in about 1927. Mass production such as this dramatically reduced the cost of cars.

Onward and upward

As we move further into the 21st century, transportation and mobility is continuing to change. One of the most recent revolutions has taken place in the skies. Aircraft have become more efficient, leading to dramatic reductions in the cost of flying. More people are flying today than at any time in history. Short-haul flights have become very available and many are offered at budget prices, making flying an affordable transportation option even over short distances. One budget airline saw its annual passenger numbers increase from just 30,000 in 1995 to around 11.4 million in 2002. As more airlines offer cheap short-haul flights, it is likely that flying will play an increasingly important role in the ongoing revolutions in transportation.

Between 1908 and 1927, Ford built over 15 million Model-Ts. Mass production techniques reduced the price of the Model-T from $850 in 1908 to less than $300 by 1925.

Progress in transportation has brought great benefits to the world, but how equally are those benefits shared by the entire world's population?

Passengers board a Ryanair plane at Manchester Airport in Britain. Ryanair is one of Europe's largest discount airlines.

KING OF THE ROAD

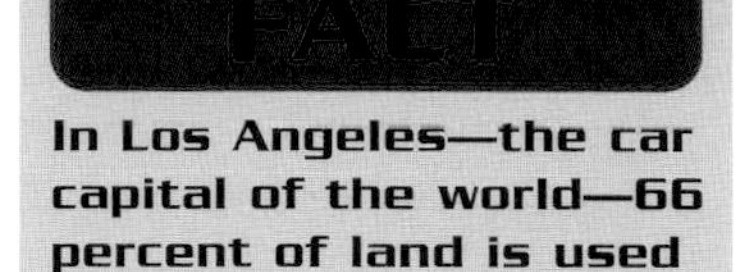

In Los Angeles—the car capital of the world—66 percent of land is used for roads and parking lots.

A motorized world

It is a measure of the success of the motor vehicle that there are few places left in the world where they are not found. In its numerous shapes and forms, including cars, trucks, and buses, the motor vehicle has become the world's most preferred method of transportation for both people and cargo. In fact, it is so favored that today's towns and cities are often planned around the needs of the motor vehicle rather than the needs of the people who live there. Between 20 and 25 percent of the land area in most towns and cities is set aside as roads and parking lots.

This highway junction in California is part of an urban landscape increasingly dominated by roads and parking lots.

The dominance of motor vehicles is spreading beyond urban areas, too. Retail parks, industrial estates, leisure complexes, and new housing developments are increasingly built in out-of-town locations. The greater use of motor vehicles perpetuates this "urban sprawl," as it is often called. As road networks expand to service urban sprawl, valuable farmland and scenic countryside are swallowed in their path. In any one location the change may seem small but, when examined at a regional, national, or even international scale, it becomes clear that we are creating an increasingly motorized world.

Good for the economy?

Politicians often support the growth of motor vehicle networks because they believe they are good for the economy and will generate more trade and employment. There is little doubt that many towns, regions, and whole countries have benefited from improvements in road networks. In East Africa, for example, many of the more prosperous settlements are located along the Trans-African Highway that passes directly through the region. Just a few miles away from the highway, levels of poverty are much higher.

The increased income brought by road networks leads to increased ownership of and dependency on motorized transporation. In time this results in roads becoming congested with too much traffic. This can harm the economy, as time and money are wasted in traffic jams and delays. In the United States, up to two billion hours are wasted each year by urban residents sitting in traffic jams. The annual cost of traffic congestion to Britain's economy is estimated to be at least $16 billion.

VIEWPOINTS

"We believe that everyone involved in transportation policy should recognize that personal mobility is an essential tool for promoting economic growth and a vital component of today's busy lifestyles."
Ford Motor Company website

"We should ... do away with the simplistic notion that 'more cars' equals 'more economic growth.'"
Federico Mayor and Jérôme Bindé, The World Ahead, *2001*

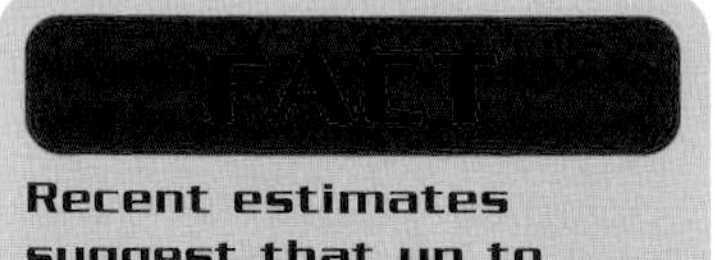

Recent estimates suggest that up to 880,000 people were killed in road accidents around the world in 1999. A further 23 to 34 million were injured.

An unhealthy habit

Our increasing dependence on motor vehicles is an issue of growing concern for human health. Nearly all motor vehicles run on petroleum-based fuels, such as gas and diesel. As these fuels are burned, a variety of waste gases and particles are released, some of which are harmful to human health. In urban areas, where traffic levels are at their highest, the problems are particularly severe, especially for those who suffer from breathing disorders such as asthma or bronchitis.

Emissions from motor vehicles contribute to this mid-afternoon smog in Beijing, China.

Among the most dangerous motor-related pollutants is lead, which is added to gas to improve vehicle performance. When released into the atmosphere, even in relatively small quantities, it can cause lead poisoning in humans. The effects of lead poisoning can include headaches, stomach pains, tremors and, in severe cases, even death. Lead is especially harmful to brain development in young children. Since the 1970s, lead has gradually been removed from fuels in North America and Europe, but it is still used in many less developed countries. In Cairo, Egypt, levels of atmospheric lead in the more congested areas are up to six times higher than those considered safe by the World Health Organization (WHO). Across the African continent, up to 90 percent of children living in major urban centers are thought to suffer the effects of lead poisoning.

Cleaning the air

In an effort to reduce air pollution from motor vehicles, governments now have numerous regulations to limit emissions and clean up the air. For example, motor vehicles in the United States and Britain must pass increasingly strict emissions tests to be allowed on the road. In Britain, poor vehicle maintenance and inaccurate testing accounts for the fact that 10 to 20 percent of vehicles operating on the nation's roads fail to meet emissions standards. However, cleaner fuels such as low-sulphur gas and diesel are helping the situation. These fuels have decreased sulphur emissions by around 66 percent from gas vehicles and 90 percent from diesel vehicles. By 2000 all diesel vehicles in Britain were using ultra low-sulphur diesel.

VIEWPOINT

"...clearly, global warming is a real issue. And even if you don't believe in climate change, there are other problems. Travel to places like Beijing and Mexico City and you can almost cut the air with a knife."

Bill Ford, Chairman of the Board, Ford Motor Company

Many authorities around the world have introduced random vehicle emissions testing to try to limit air pollution caused by motor vehicles. This roadside testing station is in London.

In Bangkok, Thailand, 524 cars were being added to the city's roads every day in 1990. By 2000 this figure had decreased slightly but was still an incredible 481 per day.

Road to ruin

The impact of vehicle emissions on human health has decreased dramatically since the 1980s. But some of the most significant waste gases, including nitrogen oxides (NOx), sulphur dioxide (SO_2), and carbon dioxide (CO_2), continue to be released in substantial quantities. NOx and SO_2 are major causes of acid rain, while CO_2 is the main gas responsible for the warming of the atmosphere and the predicted changes in world climate. In 1999 road transportation accounted for 16.9 percent of total CO_2 emissions, the same as from all forms of transportation in 1971.

Acid rain has destroyed these trees near Litvinov, in the Czech Republic.

The proportion of global CO_2 emissions from motor vehicles will increase further as the number of vehicles on the world's roads continues to grow. Much of this future increase will come from less developed countries where car use is growing most rapidly. In South Korea, for example, the number of motor vehicles on the nation's roads increased by an incredible 1,710 percent between 1980 and 1996. This massive growth led to an increase in motor vehicle CO_2 emissions of 2,183 percent over the same period.

VIEWPOINT

"Advances in transportation technology have brought benefits, but growing vehicle fleets and escalating fuel use have also created problems."
Molly O'Meara Sheehan, Worldwatch Institute

Even where the number of vehicles is increasing less rapidly, CO_2 emissions are growing disproportionately because people are traveling further than at any time in the past. In the United States, the annual distance driven by passenger cars increased four-fold between 1950 and 1999. In Britain there was a fifteen-fold increase over the same period. Such troublesome patterns have led many people to believe that motor vehicles are driving us along a road to environmental ruin.

Technology to the rescue?

Those in the motor industry point out that technological improvements over the last few decades have made vehicles more efficient and less polluting. A vehicle made today, for example, produces less than 5 percent of the emissions of its 1960s equivalent. However, while welcoming such changes, critics argue that there is room for further improvement. For instance, a European Union (EU) study found that the technology already exists to reduce fuel consumption in an average gas-powered car by around 40 percent. A similar study by the Sierra Club, a major U.S. environmental organization, found that a typical family car could be made 54 percent more fuel-efficient by using the latest technology.

VIEWPOINT

"The automobile industry's record for introducing environmental technology over the last few decades is really poor. We need the government to step in and force the automakers to build more fuel-efficient vehicles."
Jason Mark, director of the Union of Concerned Scientists (UCS) Clean Vehicles Program

A sustainable future?

Harvesting sugar cane in Cuba. Sugar cane can be used to make another form of biofuel.

The bad news is that motor vehicle numbers and use will likely continue to grow. But there is also some good news, in the form of new sustainable fuel technologies. These include biodiesel, which is made from vegetable oils and can be used in a pure form or as a blend with oil-based diesel. As a cleaner alternative to diesel, biodiesel produces fewer emissions and is renewable because it is made from plant extracts. In the United States, biodiesel was being used by over a hundred major vehicle fleets in 2002, including those of the U.S. Postal Service, the Army, the U.S. Department of Energy, and the National Aeronautics and Space Administration (NASA).

The most exciting prospect for the future of driving is the development of the hydrogen fuel cell. This cell combines hydrogen with oxygen to create electrical energy that can be used to power a vehicle. The only waste product is water vapor, so hydrogen fuel cells have the potential to be a completely sustainable and emission-free technology. Several car manufacturers are developing fuel-cell powered vehicles and the technology is already being used in trial buses in North America and Europe. The problem, according to critics, is that hydrogen fuel is currently extracted from existing fossil fuels (which are made up of hydrogen and carbon) and so is not sustainable. It is possible to produce hydrogen from water, however, by splitting it into hydrogen and oxygen using a process called electrolysis. Electrolysis itself uses a lot of energy, but if this came from renewable sources (such as wind or solar power) then hydrogen fuel cells could be used to power zero-emission motor vehicles.

VIEWPOINT

"America can lead the world in developing clean, hydrogen-powered automobiles. With a new national commitment, our scientists and engineers will overcome obstacles to taking these cars from laboratory to showroom."

President George W. Bush, State of the Union Address, 2003

This bus is one of a new generation of zero-emission buses now on trial around the world. They use hydrogen fuel cells as their power source, a technology that has the potential to bring clean transportation to future generations.

New attitudes

No matter how clean new technology is, issues such as congestion and traffic accidents will continue to make motor vehicles a problematic means of transportation. The solution, many now believe, lies in changing people's attitudes toward transportation and encouraging them to use a variety of different options. This is easier said than done, however, since many people have become very attached to their cars and the freedom of mobility that they offer. The next four chapters look at some alternatives to motor vehicles and consider both their positive and negative contributions to the transportation debate.

Does technological progress, such as the development of improved, cleaner fuels, mean that we can ignore the problems caused by motor vehicle use?

TAKING TO THE AIR

VIEWPOINT

"For nearly five decades air [transportation] has provided significant public benefits. It has brought work, prosperity, increased trade, and new travel and tourism opportunities."
Air Transport Action Group (ATAG), Switzerland

Up, up, and away!

As we move further into the 21st century, one of the fastest-growing transportation sectors is air travel. Advances in technology allow aircraft today to be bigger and to carry far more passengers than in the past. The largest aircraft can now accommodate up to 550 passengers. They have also become more fuel-efficient due to better engine design, improved aviation fuels, and the use of lightweight construction materials. Passenger jets produced in 2002, for example, used about three times less fuel per seat-mile than jets built in the 1960s. This remarkable efficiency improvement of around 66 percent compares with gains of between 15 and 30 percent for road and railroad transportation.

As a result of all these technological improvements, jets can fly farther than at any time in the past. The latest Boeing 747-400 (popularly known as "Jumbo") passenger jets have an incredible range of 8,827 miles (14,205 kilometers). This enables them to fly nonstop from New York to Hong Kong or from London to Singapore.

A 747 passenger jet, one of several long-range aircraft that have transformed international air travel.

Even more dramatic than the improvements in aviation technology over the last 50 years has been the growth in actual air travel. By 2000 the total number of passengers using air transportation for business and tourism had reached around 1.6 billion. This number is expected to increase further to an amazing 2.3 billion passengers a year by 2010. The majority of these passengers are from the more developed regions of the world where air travel is more established. For instance, Europe accounts for around 58 percent of all air passengers, while the United States has one of the most established air transportation networks. Its commercial air fleet made over 8.8 million scheduled departures in 2001.

Global air transportation of cargo is growing even faster than that of passengers, at around 11 percent a year since 1960. In 2001 about 32 million tons (29 million metric tons) of freight cargo (which represented 40 percent of the world's manufactured exports by value) was transported by air.

VIEWPOINT

"Where flying was once a privilege, even for the [rich], it's now a given in the lives of the vast majority of [British] citizens."

Jonathan Porritt, program director, Forum for the Future, Britain

FACT

In 2000 Hartsfield International Airport in Atlanta, Georgia, was the busiest in the world, handling 915,657 flights and 80.2 million passengers.

Airfreight sits on the tarmac in Hong Kong before its journey. Airfreight is a fast-growing transportation sector.

VIEWPOINTS

"The costs of failing to expand [airport] capacity for the traveling public will be felt through more delays, less choice in routes and connecting traffic, higher fares, and fewer regional air links."
Brenda Dean, Chair, Freedom to Fly

"People who live close to airports suffer more than mere annoyance from ascending and descending aircraft. Beyond annoyance, aircraft noise may have significant mental and physical health impacts on people who live below the flight path of commercial and private airplanes."
League for the Hard of Hearing, New York

FACT

Air tickets would cost 20 to 30 percent more for short-haul flights and 5 percent more for long-haul flights if the environmental costs of emissions and noise were included in the price.

High-level polluters

Air transportation currently produces around two percent of global CO_2 emissions from human sources. However, if air traffic continues to increase as expected, emission levels will triple by 2050. Emissions from aircraft are particularly damaging because they are released at high altitude, where they are known to do more damage than those released at ground level. The airline industry is constantly improving the efficiency of aircraft (and therefore reducing emissions) since fuel is one of its major costs, but the growing number of aircraft in use means the total volume of emissions continues to grow. Long-haul flights are less polluting than short-haul flights, because takeoffs are the most polluting part of the journey. However, with falling airfares, short-haul flights have become one of the fastest-growing flight sectors, especially in Europe.

To meet the increasing demand for short-haul flights, many local and regional airports are being expanded and new ones are being opened. Such developments, while sometimes benefiting the local economy, often irritate local residents who fear an increase in local air and noise pollution. Although airports are inevitably noisy places, careful planning of their location and improved aircraft design can help reduce the problem. Aircraft used today are about 75 percent quieter than those made in 1970.

For the good of people

Besides economic benefits, air transportation has brought significant humanitarian gains too. For example, in Australia the "flying doctor" service provides medical care for remote communities living beyond the drivable range of a hospital or clinic. In 2000 the service's 45 aircraft flew more than 2.8 million miles (4.5 million kilometers), and visited some 185,000 patients. Aircraft are

also used for humanitarian aid missions, such as getting food or medical supplies to areas struck by natural disasters. Air transportation enables emergency services to respond much more rapidly than they could using land- or water-based forms of transportation. And, in doing so, it saves thousands of lives every year.

Aircraft allow the "flying doctor" service in Australia to reach even the most remote communities and deliver vital healthcare services to thousands of people.

Year	Percentage increase in global passenger miles flown since 1990
2000	36
2015	115*
2030	239*
2050	524*

Source: Air Transport Action Group (ATAG) and United Nations Environment Program (UNEP), 2002

** Predicted increases*

Should we be building more airports to encourage the continued rapid growth of air transportation?

VIEWPOINT

"Ore freight on rail is a worthwhile objective. It is good for Great Britain ... for our environment, and for the [transportation] system on which our individual way of life depends."
Chairman, British Strategic Rail Authority (SRA), 2001

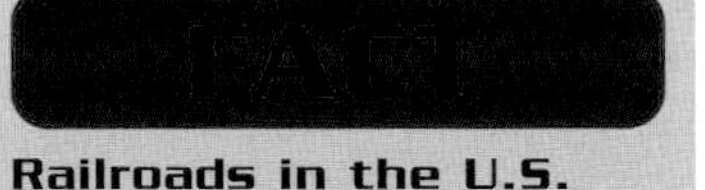

Railroads in the U.S. move more than four times as much freight as do all of Western Europe's freight railroads combined.

Vital link

Rail networks remain vital for many industries that need to move large quantities of bulky or heavy items. In the mining industry, for instance, trains are frequently used to transport raw materials to processing plants or to ports for export. In Jamaica the only railroad still running today is a privately owned line used to transport bauxite (needed to make aluminum) from the island's bauxite mines. In Britain the railroad remains the main mode of transportation used to deliver coal to the country's coal-fired power stations, which provided around 33.5 percent of British electricity in 2001.

As roads become more congested in the future, many railroad networks expect to see an increase in demand for freight transportation. In Britain the government hopes to increase rail freight by 80 percent between 2000 and 2010. If achieved, this increase would be a significant factor in easing congestion on the roads. (An average freight train can replace the equivalent of fifty semitrailer trucks on the road network.) In addition, rail freight produces around 80 percent fewer CO_2 emissions than trucks for each ton carried, and has been shown to be 27 times safer than transporting goods by truck.

This Australian freight train can carry heavy goods across vast distances cheaply, safely, and efficiently.

Greater flexibility

Many companies are unwilling to use rail freight because they believe it is inflexible. Freight companies argue, however, that improved links with road transportation will help develop a 21st-century door-to-door service for rail freight. A central part of this is the development of containerization. This involves the use of standard-sized containers for the transportation of freight, which would allow easy transfer between different modes of transportation such as ships, trains, aircraft, and trucks.

VIEWPOINT

"To survive and prosper, local and regional railroads cannot just maintain but must increase the market for railroad transportation— a challenge for the upcoming century."
Dave B. Clarke, "Local and Regional Rail Freight Transport," 2000

These containers, waiting to be unloaded from ships at a container port in Hamburg, Germany, can be easily transferred to trucks or trains.

In the United States this kind of intermodal freight transportation involving the railway almost tripled between 1980 and 2000, with 69 percent of freight being transported in containers. New technological developments, such as roadrailers, are likely to increase intermodal use of the railroads even further. Roadrailers look like normal semitrailers, but can ride directly on train tracks by using an extra set of steel rail wheels and raising the road wheels off the ground during rail transit.

VIEWPOINTS

"Light rail has been a success in virtually every U.S. city which has tried it. This is encouraging yet more U.S. cities to follow the light rail trail."
David Briginshaw, Associate Editor, International Railway Journal

"Experience in the U.S. has been that the introduction of a light rail system takes relatively few cars off the road. The problem is that freer streets tends to attract car users again."
Richard Buckley, Understanding Global Issues: Avoiding Gridlock, *1997*

Vacation Trains

Some railroad lines that have fallen out of commercial passenger or freight use are today finding new life as vacation attractions. The Grand Canyon Railway in Arizona, which was first established in 1901 to serve local mines, is now a popular tourist line. Today it takes almost 170,000 visitors a year from the town of Williams to the rim of the Grand Canyon—one of the world's greatest natural wonders. In California the Napa Valley Wine Train is equally popular as an alternative way to view the region's famous vineyards. Elsewhere in the world, trains such as the luxurious Orient Express in Europe and South Africa's Blue Train preserve memories of a bygone era when trains provided the most elegant, stylish way to travel.

A steam engine on the Grand Canyon Railway in Arizona, once used for the mining industry, is today preserved as a tourist attraction.

New direction

This light rail system in central Sydney, Australia, provides a quick and convenient method of transportation and an interesting view of the city below.

These days, the fastest-growing forms of rail transportation are light rail systems, which are increasingly being used for short-distance travel in urban centers around the world. Light rail uses tracks that are either sunken into existing roads and streets or run on elevated sections above street level.

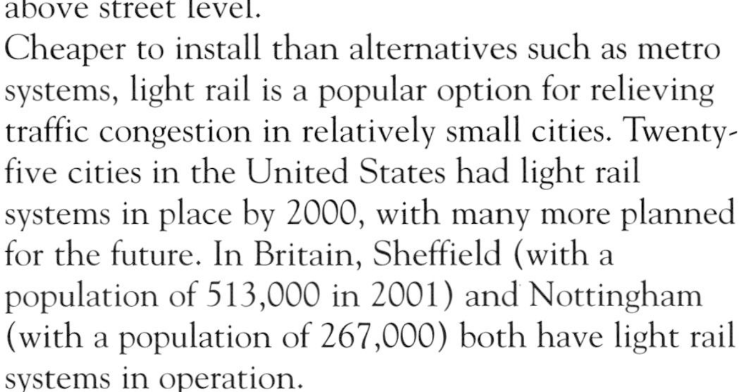

Cheaper to install than alternatives such as metro systems, light rail is a popular option for relieving traffic congestion in relatively small cities. Twenty-five cities in the United States had light rail systems in place by 2000, with many more planned for the future. In Britain, Sheffield (with a population of 513,000 in 2001) and Nottingham (with a population of 267,000) both have light rail systems in operation.

Advocates of light rail systems point out that they are highly effective at reducing car journeys and regenerating urban centers. St. Louis' light rail link carried about 42,000 passengers every weekday in 2001—80 percent of whom used to travel by car. Meanwhile, the Metropolitan Area Express (MAX) light rail in Portland, Oregon, has helped regenerate the city center and eliminated the need to build eight multilevel parking structures and widen roads linking the city to the suburbs. Critics of light rail believe that it simply creates road space that is quickly filled by more drivers. Despite these concerns there is little doubt that light rail continues to improve mobility for thousands of urban and suburban residents around the world.

DEBATE

Should more city center roads be converted to run light rail systems? What reasons would you give to support your argument?

WATERWAYS

VIEWPOINTS

"[U.S.] waterways and ports link 152,000 miles (244,600 kilometers) of rail ... and 45,000 miles (72,400 kilometers) of interstate highways. Vessels and vehicles transport goods and people throughout the system."
U.S. Marine Transportation System's report to Congress, 1999

"Unlike continental Europe, much of [Britain's] inland waterways system is unsuited to carrying significant volumes of freight. Most canals have not changed significantly from when they were built around 200 years ago."
British Department for Environment, Food & Rural Affairs (DEFRA), "Waterways for Tomorrow," 2000

Liquid asset

Water has a long history as an important means of transportation, and many of the world's greatest settlements—including New York, London, Cape Town, Rio de Janeiro, Shanghai, and Tokyo—are located on waterways or by the coast. They have developed there because of the economic benefits offered by waterborne transportation. Despite the success of trains, cars, and aircraft, this form of transportation remains very important today.

There are several reasons for this. First, no other form of transportation is able to match the carrying capacity of cargo vessels. Rivers and canals continue to perform a vital role in connecting inland areas with the coast. The Mississippi River transports about 520 million tons (472 million metric tons) of cargo each year, including almost half of U.S. grain exports. Although transportation by water is much slower than other forms, this is more than made up for by the amount that can be carried. A standard tow barge with fifteen barges can transport the equivalent of 870 semi trucks. If travelling in a convoy, spaced 150 feet (46 meters) apart, that number of trucks would stretch for 35 miles (56 kilometers).

Second, there is the question of cost. Studies published in 2002 showed that transporting freight by river was up to five times cheaper than transporting it by road. Supporters of inland water transportation claim that more businesses could take advantage of waterways if they simply planned their freight movements further in advance. In a world moving at an increasingly faster pace, however, many businesses see inland waterways as an outdated mode of transportation.

The port of Tokyo handled 40 million tons (36.4 million metric tons) of container cargo in 2000.

Oceans apart

International water transportation is an essential part of the modern global economy. Raw materials and finished goods are carried great distances across the globe in this way. Foods, including some of the tropical vegetables and fruits are now increasingly shipped over great distances and regularly made available to shoppers in colder climates in places like North America and Europe. Large ports such as Rotterdam in the Netherlands and Shanghai in China have developed on the basis of this trade. Rotterdam alone handled 355 million tons (322 million metric tons) of freight in 2002.

FACT

The largest cargo vessel in the world is the *Jahre Viking* oil tanker. It is 1,500 feet (458 meters) long and weighs about as much as 374,000 typical family cars.

The yachts moored at Causeway Bay in Hong Kong show that links between road and ocean transportation are important for leisure as well as for trade and industry.

VIEWPOINT

"There are hundreds of older vessels in the world fleet that are simply ticking time bombs. The fact is they shouldn't be allowed to carry toxic cargoes, never mind pass anywhere near pristine coastlines."
Anonymous European salvage expert, Planet Ark news service

Passenger traffic

Besides freight, water transportation is also used to carry people for business, leisure, and tourism purposes. In the Swedish capital of Stockholm, a city divided by numerous rivers and channels, passenger ferries are a regular part of the city's public transportation system. For longer-distance travel—between Britain and Europe across the English Channel, for example—larger ferries (which also carry vehicles) are one of the most common forms of transportation. Dover, Britain's busiest port, was used by some 16 million passengers and 2.5 million cars and buses in 2002. Longer journeys, such as transatlantic crossings, are today mostly taken by tourists as part of the rapidly growing cruise industry. And in countries such as France, Britain, and the Netherlands, many people enjoy taking leisurely boating vacations and trips on rivers and canals.

Riverboats wait for their passengers on the Volga River in Russia.

Disrupting nature

However, there are some downsides to water transportation. For efficient operatation, considerable changes often have to be made to the surrounding environment. Rivers are straightened, dredged, redirected, or dammed while coastal environments are cleared to make way for big ports. The adaptation of rivers to encourage transportation can be particularly disruptive to natural environments and processes. It can change the flow of water and, in times of high flow, lead to devastating flooding as the river tries to regain its natural course. The Yangtze in China regularly suffers from such flooding. And the Mississippi river burst through several levees (artificial embankments) in 1993, causing $12 billion worth of damage and stopping river transportation for two months.

During the 20th century, more than 200 oil tankers have sunk—many of them resulting in ecological disasters.

Improving rivers for transportation often requires protective measures such as levees (earth embankments). This aerial view of the Atchafalaya River in Louisiana shows the high levees that have been built along the river basin.

Other environmental impacts include the sometimes serious damage boats can cause to aquatic species through accidents and pollution. In India, populations of the now endangered Ganges river dolphin have decreased dramatically because of accidents with boat propellers and pollution from boat engines. Spills from the large oil tankers that transport fuel around the world are perhaps the biggest environmental hazard of water transportation. When the *Exxon Valdez* ran aground off Alaska in 1989 it spilled 257,000 barrels of oil along 200 miles (320 kilometers) of Alaskan coastline. This caused the deaths of an estimated 350,000 seabirds.

FACT

Each day, more than 6,000 oil tankers are roaming the seas and oceans.

Responsible shipping

Following accidents such as the *Exxon Valdez*, and many that have happened since, new international standards have been agreed upon to phase out the older and more hazardous ships currently using the world's waterways. By 2015 all oil tankers will have to be double-hulled instead of single-hulled ships such as the *Exxon Valdez*. This will ensure that their cargo is better protected and will make spillage less likely if they hit submerged objects or are involved in a collision. However, it may prove difficult to enforce such regulations, and some single-hulled vessels will probably continue to operate illegally beyond the 2015 deadline.

VIEWPOINT

"As soon as there is an incident like this [the sinking of the *Prestige*] everyone throws up their hands and tries to deny responsibility—that really has to change. The shipping world still has a long way to go in terms of governability."
David Santillo, Greenpeace International

One of the biggest problems in enforcing shipping regulations is the fact that it is such an international industry. Take the example of the oil tanker *Prestige,* which sank off Spain in November 2002. It released 11,000 tons (10,000 metric tons) of oil and took another 72,000 tons (65,000 metric tons) with it to the bottom of the ocean. At the time the *Prestige* was carrying oil from Latvia to Singapore for a Russian trading company based in Switzerland. The ship itself was Liberian-owned, but registered in the Bahamas. When it sank, it was operated by a Greek-based company and had been passed as seaworthy by an American shipping authority. With so many different groups involved, it can be difficult to decide who should take responsibility when things go wrong and who should be responsible for putting them right.

VIEWPOINT

"If the oil tanker [*Prestige*] loses all its oil ... if all that escapes from the hull, then this is a disaster which is going to have twice the effect of the *Exxon Valdez*, which is one of the worst that we have known."
Christopher Hails, World Wide Fund for Nature [WWF], discussing the Prestige *oil spill off Spain, November 2002*

Following the *Prestige* sinking, many environmentalists and politicians have called for the ban on single-hulled vessels to be moved forward. According to some reports, the ban could be achieved by 2010–2012, but many shipping companies say that moving the deadline forward is simply not possible. They say that time is needed

Volunteers clean up the oil spilled by the tanker Prestige *on the Spanish coast in November 2002. Such incidents are a dramatic reminder of the risks involved in transporting hazardous cargo on the ocean.*

for the massive investment and construction process required to build new double-hulled oil tankers. There are fewer shipyards able to build such vessels than in the past and there are concerns that rushing the new regulations could lead to poor-quality vessels and increase the chances of a disaster in the future.

DEBATE

Given the risks involved in transporting oil by water, is our dependence on oil as the main fuel still a sensible policy in the 21st century?

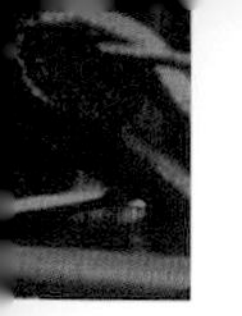

RECLAIMING THE STREETS

FACT

In Uganda 41 percent lof households have a bicycle. By comparison there are just 1.8 passenger cars for every 1,000 people.

Back to basics

In many towns and cities across the world, communities are deciding that they have had enough of the problems brought about by existing transportation options. They have a right to be concerned. There are many problems, including city air thick with pollution, buses delayed by traffic jams, roads that are too dangerous to bike on, and priority given to cars instead of pedestrians. But there is a growing demand to rethink urban transportation and, in many cases, go back to basics by encouraging a revival of nonmotorized modes of transportation, such as walking and biking.

Nothing new

Of course there is nothing new about walking or biking. Walking is humankind's most ancient and natural method of moving from one place to another, and is still the dominant form of mobility for much of the world's population. In less-developed countries (where the vast majority of the world's population lives), people often have little choice but to walk. In villages throughout sub-Saharan Africa, it is common for people to walk for several hours a day just to perform basic tasks such as collecting water or fuel wood. In addition to the time and distance involved, this places an enormous physical burden on those who regularly carry loads of 44 pounds (20 kilograms) or more. In these situations alternatives to walking are required to make people's daily lives easier and to help the development of local economies.

VIEWPOINT

"The bicycle's principal attraction is its low cost. With cars costing easily 100 times as much, the bicycle offers mobility to billions of people who cannot afford a car."
Lester R. Brown and Janet Larsen, Earth Policy Institute, Eco-Economy Updates 13, 2002

With little alternative, these Kenyan women regularly carry heavy loads on their heads.

In North America and Europe, the bicycle is often considered a rather old-fashioned and basic form of transportation, reserved today for recreation and leisure. In the United States the number of young people between seven and seventeen years old who rode a bicycle more than once a year fell by 13 percent between 1990 and 1999. In stark contrast, bicycles are seen as prized possessions in many less developed countries. They greatly improve mobility for those who are unable to afford other transportation, allowing easier access to markets and reducing the burden of carrying heavy loads on foot. Bicycles can also help in the provision of essential services such as health and education. In Ghana, West Africa, bicycles given to HIV/AIDS educators have helped them communicate their vital message to 50 percent more people than have been reached by those educators working on foot.

VIEWPOINT

"...cycling is still often linked in the public mind to a society that pre-dates [wealth] and prosperity. The use of bicycles must be transformed into a point of pride, with the emphasis on its advantages as a silent, flexible, nonpolluting means of transport, excellent for health and requiring little infrastructure."

Federico Mayor and Jérôme Bindé, The World Ahead, *2001*

VIEWPOINTS

"Cities can actively promote walking and cycling by investing in bike paths and racks, slowing cars, and making streets physically appealing. Studies of cycling investments in Amsterdam [the Netherlands], Bogotá [Colombia], Morogoro [Tanzania], and Delhi [India], found that small investments can yield great benefits."
Molly O'Meara Sheehan, Research Associate, Worldwatch Institute

"Bikes may be fun on a sunny afternoon, but who wants to cycle in winter rain? Problems of weather-protection, security, and baggage capacity, all put cycling at a severe disadvantage, despite its environmental benefits."
Richard Buckley, Understanding Global Issues: Avoiding Gridlock, *1997*

Encouraging signs

Recent trends suggest that bicycle use is increasing. In 2000, 101 million bicycles were produced worldwide (more than twice the number of motor vehicles and the highest production figures since 1995). At the same time, the infrastructure needed to encourage greater cycling is receiving a welcome boost in many countries. An extensive network of cycle paths and cycle-safe routes was opened in Britain in 2000 as part of the National Cycle Network. In 2002 an estimated 97.2 million trips were made on the 6,750-mile (10,860-kilometer) network, a figure that is expected to grow as the network extends to 10,000 miles (16,000 kilometers) by 2005.

Practical solutions

One of the biggest barriers to cycling is that it is not always a practical transportation solution. Those living in hilly areas, for example, may find it too difficult to bike regularly, while some elderly people may find even cycling on level ground too much of a strain. New developments, such as electric bicycles (e-bicycles), can help solve such

Policemen use bicycles on duty in Venice Beach, California.

problems. In 2000 the world sales of e-bicycles reached 1.1 million—three times the number sold in 1999. E-bicycles are especially popular in China where they are regularly used by commuters. They are also gaining in popularity in the United States where several police forces now use them in cities like Chicago and Dallas. Officers on bicycles have been shown to reach crime scenes quicker than those struggling through traffic in squad cars and typically make 50 percent more arrests.

China, India, and Taiwan produced 65 million bicycles in 1999, over half of which were exported.

In central Copenhagen, Denmark, free bikes are provided for use by members of the public.

If cycling is to be further encouraged, facilities such as secure parking for bicycles and cycle lock-ups at stations, colleges, and places of work will also need improving. Where this has been done in combination with developing cycle routes the use of bicycles has increased. In Copenhagen, Denmark, for example, about 30 percent of commuters travel to work by bicycle, while in the Netherlands up to half of city journeys are made by bicycle. Even in the United States,where cycling is on the decline, an estimated 3 million people bike to work each day.

An e-bicycle can usually travel at speeds of up to 18 miles (29 kilometers) per hour, unassisted by pedaling.

It makes sense

Making walking and cycling easier and more appealing not only helps improve the quality of urban environments and people's health, it can also make good economic sense. Biking, in particular, is often the quickest and most efficient way to move around congested urban streets. In London a 1996 government survey found that a 1.7-mile (2.7-kilometer) journey took cyclists an average of just 18 minutes to complete, compared to 33 minutes for those traveling by car and 38 minutes by bus.

A bicycle courier cycles down Fifth Avenue in New York City. Bicycles can be the fastest way to move goods around increasingly congested city streets in many parts of the world.

Courier companies are increasingly taking advantage of the greater mobility bicycles allow and employing bike messengers to travel busy city streets. In New York about 300 bicycle messenger companies compete for $700 million worth of business each year. In Sydney bicycle couriers have been criticized for endangering the safety of others by riding on sidewalks, jumping traffic lights, and ignoring one-way streets. Complaints led the city council to introduce new regulations for bicycle couriers since 2003.

Home zones

Although pedestrianized city centers may be a familiar sight today, it is less common to see whole streets that have been made more people-friendly. This is beginning to change, however, with the spread of "home zones." Home zones involve traffic-calming measures (such as speed bumps), speed limits of 10–20 miles (16–32 kilometers) per hour, and changes of priority so that drivers must give way to pedestrians and cyclists. In addition, home zones are often planted with trees and flower beds and may have benches, play areas, or works of street art. In the Netherlands home zones are extremely popular, with over 6,500 now in place across the country. Germany, Austria, and Denmark have also adopted the home zones idea and it is currently being tested in the United States and Britain.

Despite the benefits they offer, home zones have met with some resistance. In the United States residents have asked for traffic restrictions to be removed because they can no longer park their cars outside their homes. In Britain there have been complaints about traffic backups and slower response times for emergency vehicles. It seems that home zones may only succeed if they are part of bigger changes in transportation policy.

VIEWPOINTS

"Historically streets were places where people lived, met, played, traded, and traveled. They were busy places: the heart of the community. But today streets are congested with traffic rather than busy people. They are seen as a source of danger."

Transport 2000 Trust, Britain

"Home zones are streets where you can enjoy taking time to stop and chat with your neighbors. Streets where cars are allowed, but the car driver is a guest. They are common and popular in many European countries."

Home Zone News, Children's Play Council, Britain

DEBATE

Would you like to see your local street made into a home zone? What would the benefits and drawbacks be for your street?

TRANSPORTATION PLANNING

FACT

On the average, about 20 percent of the time taken to complete a journey using public transport is "waiting time" at stops and interchanges.

Thinking ahead

It may seem obvious to say that transportation requires careful planning, but it is sometimes surprising how little thought goes into new projects and policies. Traffic lights, for example, are rarely timed for the benefit of pedestrians, even though it is those on foot who have to stand in the cold and rain waiting to cross while motorists whizz past in their warm, dry vehicles.

Government transportation policies often seem to be very shortsighted. For instance, the continued emphasis on creating more road capacity in many countries ignores the fact that motorized transportation is causing great harm to environments and human health. Motorized transportation is also currently dependent on oil—a nonrenewable resource. Environmentalists and an increasing number of urban planners argue that we should be looking for transportation solutions that

An enormous area of land is cleared to make way for a new highway.

reduce our dependence on the car. The challenge is to find a way of doing this without damaging the economy or reducing personal freedom or mobility.

Waiting times can discourage many people from using forms of public transportation.

Working together

The key to transportation planning is to achieve greater coordination between the different methods available. Containerization in the freight industry is a good example of this, and has significantly speeded up the movement of freight between different modes of transportation. However, moving people in a coordinated fashion can be much more complex than moving goods. Different modes of transportation follow different timetables and run on set routes, so getting from A to B can prove complicated and frustrating. For instance, delays to an intercity train service may cause passengers to miss a local train connection and extend their travel time significantly.

The same problem often occurs at a more local level with difficulties in synchronizing bus services or in sychronizing light rail systems and buses. In the Austrian city of Graz and in Singapore, public transportation has been brought under the management of a single authority to reduce such problems. There is a single transferable ticketing system, and timetables are coordinated to make life as easy as possible for passengers. These improvements have created an increase in the use of public transportation and a decrease in car use in both cities. Such systems are providing valuable lessons for other cities around the world and are now being copied in many places.

VIEWPOINTS

"The constantly increasing level of road congestion has become a major problem for the [entire] industry. It is absolutely vital the [British] government takes every possible action to improve the operation of the existing road network, as well as making longer term plans for motorway upgrades and widening."
Richard Turner, Chief Executive, Freight Transport Association, Britain

"Adding highway capacity to solve traffic congestion is like buying larger pants to deal with your weight problem."
Michael Replogle, Environmental Defense

VIEWPOINT

"Park & Ride allows commuters ... to enjoy stress-free travel and beat the [traffic]. It also allows visitors and shoppers to take the direct route to the heart of the city."
Cambridgeshire County Council website, Britain

Changing priorities

Those who campaign for improved transportation policies believe that one of the most important tasks is to change transportation priorities. In particular, they argue that private cars should be given lower priority than other forms of transportation. Many cities are already adopting such an approach by doing things such as marking out dedicated bus lanes. This helps to speed up local bus services, making them more frequent and reliable to encourage more people to use them. Bus lanes have been particularly successful when combined with "park and ride" facilities. These allow people to leave their cars at one of several parking lots around the city limits and then catch a regular bus service into the downtown area.

A section of road in Paris, France, is set aside exclusively for buses. Priority lanes such as these can greatly improve the efficiency of public transportation.

Priorities can be assigned in other ways, too. In the Netherlands buses are given priority at traffic lights thanks to an intelligent traffic control system. In the United States, high occupancy vehicle (HOV) lanes are increasingly common, allowing cars carrying at least two people to use a separate, faster-moving lane. Such measures have encouraged car-pooling programs in cities such as Los Angeles, where people meet at parking lots and then share a ride downtown using the HOV lane. Many parking lots have lock-up facilities for cyclists, so the system is not just for car users. Providing more secure storage for bicycles has been central to the growth in bicycle use in European cities such as Copenhagen, Strasbourg, and Freiburg. It is also a very sensible policy, because six bicycles can use the road space taken up by a single car and around twenty will fit into a typical car parking space.

Safe storage for bicycles, shown here in Ravenna, Italy, is essential to encouraging their greater use.

Access for all

Of course, any planning changes must ensure that everyone has access to transportation. Policies that ban motorized vehicles from urban areas and pedestrianize main streets can be a problem for the elderly or disabled who may not be very mobile. Planners must ensure that these people continue to have access and that the available transportation meets their needs, too. This is especially true of many European societies where the population is aging rapidly. In Italy an estimated 35 percent of the population will be over 65 in 2050.

In a recent survey, two-thirds of people questioned believed that pedestrians and cyclists should be given priority in towns and cities. Only one in seven disagreed.

VIEWPOINT

"The highways that are built to sustain sprawling suburbs add to our pollution and energy problems, and increase our dependence on an auto-centric way of life, which is unhealthy, antisocial, and unsustainable."
Sierra Club, "Stop Sprawl" campaign

The big picture

Some of the most successful transportation policies have been those where transportation is considered part of a wider planning process. Curitiba, one of Brazil's fastest-growing cities, is a good example of this. In the city and its suburbs, all new developments are planned in such a way that they are served by the city's highly efficient bus network. A system of feeder buses serves the suburbs and links with the downtown service that runs on dedicated bus highways to ensure a frequent and regular service. In this way all parts of the city are connected, and almost 70 percent of the population uses the bus system each day. Central to the success of Curitiba's bus system isa flat-rate fare. This avoids the problems encountered in some other cities where communities living on the city outskirts cannot afford to use public transportation. In Manila in the Philippines, for example, lower-income residents spend up to 14 percent of their income on traveling into the city to work.

This very successful and environmentally friendly bus system in Curitiba, Brazil, is now being copied in other cities across South America.

Rethinking land use

In the United States and Britain, new developments such as out-of-town shopping malls and industrial parks are increasingly spreading beyond the city center. Other countries, including the Netherlands and Japan, have banned such developments. They believe them to be unsustainable because they encourage a greater dependency on private cars and are often poorly connected to existing public transportation networks.

Some retailers have been accused of encouraging greater car use by setting up out-of-town locations, away from public transportation networks.

In the Netherlands a land planning system called "ABC" was introduced in 1990 to encourage development that suits the availability of transportation facilities. "Group A" locations, for example, are those easily accessible by public transportation and are designated for shops and offices where there is a high flow of people. Parking spaces are limited to just 10 per 100 employees to help keep car commuting below 20 percent. Greater car use is expected in "group C" locations, but parking spaces are still limited to 40 per 100 employees. Many people believe that this type of innovative urban planning is essential to the future of transportation, but such policies are difficult to enforce unless they have the support of those they are intended to benefit.

Should out-of-town developments, such as shopping malls and leisure facilities, be banned in the interests of more sustainable transportation?

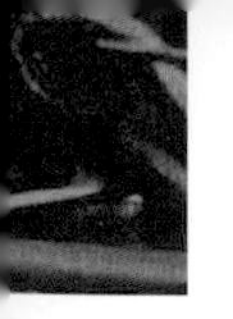

CHOICES AND ACTION

FACT

Transportation in sprawling cities such as Houston, Texas, produces almost five times more CO_2 emissions than transportation in a densely planned city such as Hong Kong.

More choice

For some areas and people, there are relatively few transportation choices. Remote rural areas, for example, may have no public transportation services at all, while cycling is probably inappropriate for most elderly people. In other regions people may be unable to afford a private motor vehicle or may not have a driver's licence. Because the transportation needs of individuals are so varied, it seems that the only way to meet everyone's requirements is to provide them with more choices.

Sparsely populated regions, like this farming area in Iowa, often suffer from a lack of public transportation services.

In the Czech Republic, public transportation use fell and car use grew as the number of suburban supermarkets increased from 1 to 53 between 1997 and 2000.

This would enable people to select the most appropriate mode of transportation for each specific journey. For a small shopping trip to a local shop, walking or cycling might be best, whereas a larger shopping trip could be better planned using the bus or light rail service. Of course some trips will be better suited to a car (if people have access to one), such as visiting relatives in a remote part of the country or traveling late at night when security may be a concern. The same sort of choices also apply to the transportation of freight, with air, rail, boat, and road each having particular benefits and drawbacks.

In general urban areas have more transportation choices because their higher population density makes transportation systems cheaper to operate. Still, there is a tremendous dependence on cars and other personal vehicles, as shown in the chart below.

A busy bus, train, and streetcar interchange at Freiburg Station in Germany. Densely populated urban centers usually have greater transportation choices than rural areas.

Old habits die hard

In several countries, including the United States and Britain, people continue to rely overwhelmingly on private cars, even when a good choice of alternative transportation exists. This is an extremely wasteful use of transportation resources. For instance, some 60–80 percent of car journeys are not essential or provide no real benefit over using available public transportation. These figures show that continuing to build a transportation system around the private car is not only bad for the environment, congestion, and public health, but is also a poor use of resources.

This chart shows that most Americans use cars rather than walking or public transportation.

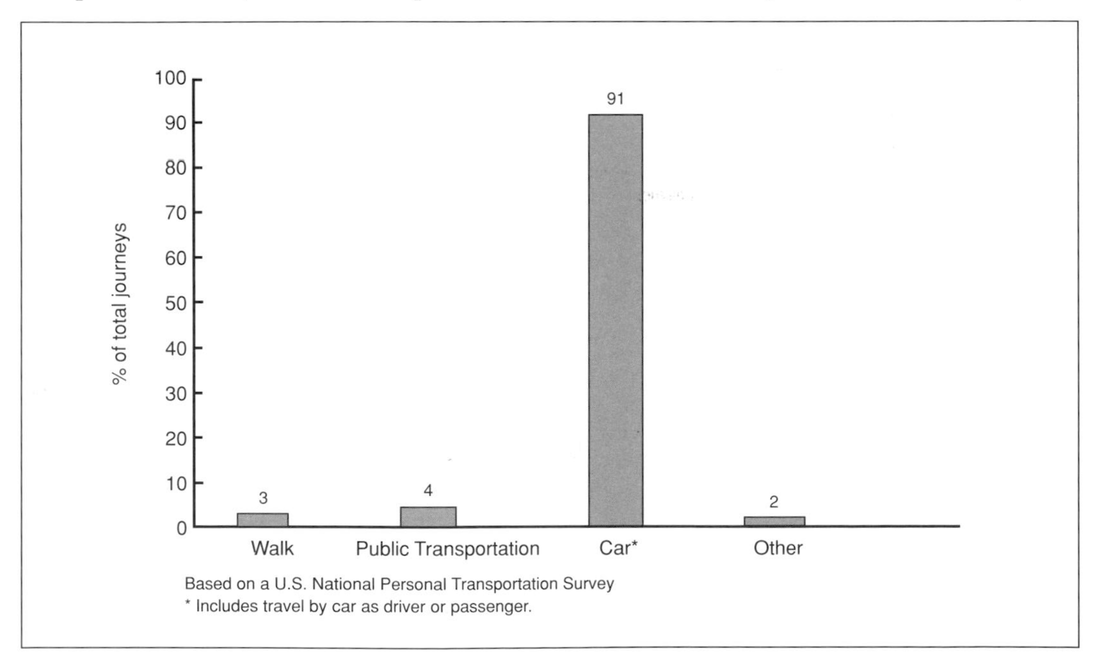

VIEWPOINTS

"We're very hopeful that car pooling will take off here because Toronto traffic is becoming more of a problem. We only have a fixed number of parking spots and the new open space program promotes more bicycle and pedestrian traffic."
Gary Nower, manager of grounds, University of Toronto, Canada

"Given the choice between car pooling and driving alone, I'd rather drive alone and pollute the air."
Participant in Washington, D.C. research project

Rising to the challenge

In several European cities, and in an increasing number of North American cities, communities have initiated voluntary responses to transportation problems. Car pooling is one example, and it operates in various ways. At the most basic level, people simply combine their journey requirements and share a car trip instead of driving individually. In U.S. cities such as Los Angeles, Seattle, and New York, car pool programs (where people park in special lots on the city outskirts and share the final part of their journey into the city) are now common.

More complicated programs exist in several cities where, instead of owning their own vehicles, people join a car share program and book use of a vehicle when they need it. This reduces their costs and means they can choose vehicles according to their needs, such as a small car for a local shopping trip or a larger "people-carrier" for a family day out. They can also choose to use a car for as little as one hour, allowing them to avoid the expense of

A sign advertises a car sharing program in Venice, Italy. Such programs are becoming more popular in Europe.

Road charges, or tolls, are another way of reducing pollution and congestion. Drivers think twice about driving their cars if they have to pay extra to use roads and highways. This toll barrier is near the popular Spanish tourist resort of Alicante.

traditional car rental companies that normally charge full-day or half-day rates. Car shares can be small or large. In Berlin a local neighborhood program shares just 3 cars among 40 neighbors. In contrast, Switzerland (one of the first countries to start car sharing) has some very large programs such as "Mobility Car Sharing" which, in early 2003, had 1,770 vehicles serving 400 communities or 49,200 customers.

New attitudes

The success of initiatives such as car sharing will depend on people adopting new attitudes to transportation. Experts argue that, far from robbing people of their mobility, greater use of public transportation and programs such as car sharing will actually increase their mobility by removing unnecessary traffic from the roads. In Switzerland, for example, studies have shown that people using a single shared car can remove the equivalent of about ten private vehicles from the roads. The problem is that many people are reluctant to share rides with strangers, or to give up their vehicles, which they believe give them ultimate freedom and mobility.

FACTS

At the beginning of 2003, there were over 36 cities in the U.S. that had started car share programs.

DEBATE

Should individuals be expected to give up their vehicles for the good of the environment and for the benefit of the communities they live in?

THE FUTURE OF TRANSPORTATION

A balancing act

Getting the future of transportation right will be a tricky balancing act. With some forms of transportation, such as long-haul air travel, there is no realistic alternative because of the time involved in such journeys. However, short-haul flights could be replaced by high-speed trains that would carry more passengers and have less environmental impact. France's high-speed TGV train between Paris and Lyons led to a 50 percent decrease in air passengers within a month when it was introduced on that route in 1981.

FACT

Online shopping in the United States increased almost 30 percent between December 2001 and December 2002.

At a more local level, although privately owned cars remain the preferred mode of transportation in many countries, evidence suggests that this is not sustainable for the future. Alternatives such as walking, biking, and public transit systems (including light rail and buses), could provide solutions, but will not be fully accepted until factors such as reliability, frequency, and safety are improved. Car users will need to see real advantages in the alternatives before they are prepared to make the switch.

Technology solutions

Modern technological developments, such as improved fuels and cleaner, more efficient engines, can help to reduce the impact of transportation on the environment, but will not solve problems like congestion and traffic accidents on their own. Technology can help fulfill these broader goals, however, by providing people with improved transportation information and making it easier to use these services. In some cities bus and train shelters now display live updated information about when the next service is due. The Internet is also being used for booking car share vehicles online or finding travel companions to establish a local car pool.

In some cases the Internet is being used to avoid journeys altogether. Online shopping is growing in popularity around the world. Many supermarkets now allow people to do their food shopping online and even deliver it free of charge. In business, too, the Internet is used for videoconferencing, allowing meetings to be held without people actually traveling to meet in person. Internet banking also eliminates the need to travel to handle financial matters in person. These options still remain limited in many less-developed countries. However, as technology costs continue to fall and more people get online, such services may help to reduce transportation demand in the future.

A teenager in Austin, Texas, uses the Internet for home shopping. Future generations could grow up taking such services for granted and reduce the need for mobility and transportation.

Force for change

Some believe that transportation problems can only be solved by direct action from local authorities or national governments. Such actions might include banning private vehicles from city centers, the raising of road or fuel taxes to discourage drivers, or the introduction of tolls. Tolls are already in place in many cities such as Singapore, which introduced tolls in 1998 to reduce the volume of traffic entering the city during rush hour. Other cities are now introducing similar systems. In Britain motor vehicles that enter central London during the day (between 7:00 A.M. and 6:30 P.M.) have had to pay a daily congestion charge since February 2003.

This scene of congested freeways in Los Angeles, California, is repeated all around the world. Do people want to face a future of endless traffic jams?

However, forcing change is often unpopular with voters, even where surveys have shown that transportation issues are considered a public priority. People want to see alternatives in place before they are forced to change their current transportation methods. But it is difficult to establish alternatives until there is enough demand to make the necessary investment worthwhile.

Personal actions count

Environmentalists and transportation activists are increasingly focusing on the message that personal actions count. They are encouraging people to reconsider their need for transportation and choose the best option for each specific journey. In Britain about 25 percent of trips are less than 2 miles (3.2 kilometers) and 58 percent are less than 5 miles (8 kilometers) long—distances that could be walked or biked by most people.

Support for rethinking transportation is certainly growing. Millions of individuals across the world participate in "car-free" days and use public transportation, walk, or bike instead. Since they began in the late 1990s, car-free days have become major events. In Europe, the event was expanded to become "European Mobility Week" in 2002, a week-long focus on different forms of transportation, each publicized on its own day. Such events certainly raise awareness of solutions to some of the world's transportation problems, but they are of little use unless those solutions are put into everyday practice. This book has shown that technology and governments may provide some of the answers, but it is ultimately up to individuals like you and me to take the problems personally and make our own transportation choices for the 21st century.

VIEWPOINT

"The ... daily congestion charge will help get London moving. It will reduce traffic, making journeys and delivery times more reliable, and raise millions each week to re-invest in London's [transportation] system."
Transport for London website

DEBATE

From what you have learned in this book, what would you select as the three most important actions that individuals could take to ensure better transportation in the 21st century? What reasons do you have for your choices?

GLOSSARY

acid rain acidic rain produced when pollutants such as sulphur dioxide and nitrogen oxides (emitted when fuels are burned) mix with water vapor in the air to form an acidic solution that falls to the ground. It is damaging to plants, trees, lakes, and buildings.

altitude measured height of an object (e.g. aircraft) or location (e.g. mountain) above sea level, normally measured in feet or meters

biodiesel motor fuel made from plant extracts that is used in a pure form or mixed with conventional diesel fuel. Biodiesel is cleaner than standard diesel fuels.

boat-plane form of aircraft that has an underside similar to a boat's hull that allows it to take off and land on water

budget airline airline company that specializes in low-cost air travel. Such flights are normally on short routes between major cities.

car pool system in which people share the use of cars in order to reduce the number of vehicles on the road, share costs, and limit the environmental impact of motor vehicle emissions

car sharing program in which people share the costs and use of communal vehicles rather than having their own. Members of a car sharing program only pay for the distance and/or time that they actually use a vehicle.

climate change process of long-term changes to the world's climate (warming or cooling). This occurs naturally, but is today occurring as a result of human activities polluting the atmosphere.

commute regular travel to work, usually from a suburb to a city

congestion when vehicles overcrowd a street or road, making movement difficult or impossible for some time. Sometimes called traffic jams.

containerization packaging of goods and freight into standard-sized containers that can then be easily transported by truck, train, or ship and transferred between different types of transportation

courier person who delivers important or urgent packages by hand. A bicycle or car is normally used for transportation.

cycle lock-up storage facility for the safekeeping of bicycles while they are not in use. They are often found at people's places of work or in public places such as train stations.

deforestation removal of trees, shrubs, and forest vegetation. This can be natural (due to forest fires, storms, etc.) or a result of human action (logging, ranching, construction, land clearance, etc.).

developed countries generally wealthier countries of the world including Europe, North America, Japan, Australia, and New Zealand. People living there are usually healthy, well educated, and often work in a wide variety of high-technology industries.

e-bicycle bicycle fitted with an electric (normally battery-powered) motor to assist the rider and reduce the need for them to pedal

emissions polluting waste products (normally gas and solid particles) released into the atmosphere. These include carbon, sulphur, and lead from car exhaust fumes.

erosion process whereby something becomes worn (eroded). For example, the removal of material (soil or rock) by the forces of nature (wind or rain) or people (deforestation, vehicle tracks, etc.).

fossil fuels fuels from the fossilized remains of plants and animals formed over millions of years. They include coal, oil, and natural gas. Once used, they are gone—nonrenewable.

freight goods carried by commercial vehicles, sometimes also known as cargo

fuel tax charge on the purchase of fuel. Such taxes can be used to change fuel consumption patterns.

global warming gradual warming of the Earth's atmosphere as a result of greenhouse gases, such as carbon dioxide and methane, trapping heat. Human activity has increased the level of these gases in the atmosphere.

gridlock situation where congestion affects a wider area than usual so that vehicles are unable to move in any direction at all

heavy goods vehicle (HGV) normally a tractor and trailer unit for transporting heavy goods by road, but more generally used to describe any cargo vehicle weighing 8 tons (7.5 metric tons) or more

high occupancy vehicle (HOV) vehicle carrying at least two (sometimes defined as three) passengers

home zone area in which traffic is limited or banned in order to create safer and more enjoyable streets for the people who live there

hull main body of a ship or airplane

hydrogen fuel alternative fuel currently being developed that releases no pollutants. It is made by combining hydrogen and oxygen in a special fuel cell.

infrastructure networks that enable communication, people, transportation, and the economy to function. These networks include roads, railroads, electricity and phone lines, and oil and water pipelines.

intercity transport which connects two or more cities.

light rail relatively small-scale (sometimes driverless) train systems that run on sunken or elevated tracks, providing a quick, efficient, and cleaner means of transportation. They are increasingly popular in busy city centers.

mass transit movement of large numbers of people

metro railway system in a town or city that runs either completely or partly underground

mobility ability to move from one location to another for leisure or work

motorized transportation any form of transportation that has a motor fitted to it, but usually restricted to mean road vehicles, such as cars, trucks, buses, and motorcycles

nonrenewable resources those that, once used, are gone and cannot be replaced. These include coal, oil, and natural gas.

park and ride program where cars are parked in out-of-town parking lots and drivers travel into urban centers on buses, trains, or streetcars, therefore reducing congestion and pollution

public transit system transportation system, such as a bus or metro network, designed to meet the mobility needs of the public

public transportation network of passenger vehicles, such as buses, trains, and streetcars, running on set routes with set times and fares

regenerate (a town center) process designed to bring new growth to an area following a period of decline. Such plans often involve improved transportation to make the area more attractive to businesses and shoppers.

roadrailer road trailer that has been adapted to ride directly on rail lines by using an extra set of steel wheels and retracting its road wheels

seat-mile the movement of one seat space (on an aircraft, for example) over a distance of 1 mile. If an aircraft has 100 seats and travels 100 miles then it has completed 10,000 seat miles (100 x 100 = 10,000).

short-haul flight flights to destinations which are a relatively short distance away

smog mixture of fog, smoke, and airborne pollutants such as exhaust fumes

streetcar passenger coach that runs on metal rails sunken into existing roads or on special tracks. Streetcars are normally powered by electricity from overhead cables. Known as trams in Europe.

suburb district, normally dominated by residential homes, located on the outskirts of a major urban center

subway train system in a town or city that runs underground

sustainable transportation alternative transportation systems that meet the needs of today's global population without causing harm to people or the environment, now or in the future

synchronize (different transport systems) to coordinate timetables so that different transportation services connect with each other. Synchronizing a long-distance train with local train or bus services, for example, would mean that passengers did not have to wait for long periods in between journeys.

traffic calming set of measures designed to slow the flow of road traffic in order to make roads safer for cyclists and pedestrians

traffic control system system (normally computerized) designed to control the flow of traffic. Such systems can be used to give priority to public transportation such as buses and trains.

urban sprawl process by which new urban development (e.g. housing, leisure, and retail complexes) spreads beyond existing town or city boundaries onto surrounding land

zero emission vehicle vehicle, such as a bicycle, that does not release any polluting waste products

FURTHER READING

Church, Amanda, and Andrew Church. *Transportation*. Chicago: Raintree, 1999.

Dils, Tracey E. *The Exxon Valdez*. Broomall, Penn.: Chelsea House, 2001.

Hall, Margaret. *Transportation*. Chicago: Heinemann, 2001.

Ray, Kurt. *New Roads, Canals, and Railroads in Early 19th-Century America: The Transportation Revolution*. New York: Rosen, 2003.

Sandler, Martin W. *On the Waters of the USA: Ships and Boats in American Life*. New York: Oxford University Press, 2003.

Sandler, Martin W. *Straphanging in the U.S.A.: Trolleys and Subways in American Life*. New York: Oxford University Press, 2003.

USEFUL ADDRESSES

Coalition for Appropriate Transportation
915 Hamilton Street
Allentown, PA 18101
Tel: 610-434-9100

Environmental Protection Agency (EPA)
Office of Transportation and Air Quality
1200 Pennsylvania Avenue, NW
Washington, D.C. 20460
Tel: 734-214-4333

Friends of the Earth
1025 Vermont Ave. NW,
3rd Floor
Washington, D.C. 20005-6303
Tel: 202-783-7400

Institute for Transportation and Development Policy (ITDP)
115 West 30th Street
Suite 1205
New York, NY 10001
Tel: 212-629-8001

U.S. Department of Transportation
400 7th Street, S.W.
Washington, D.C. 20590
Tel: 202-366-4000

INDEX

Numbers in bold refer to illustrations.

INDEX